Zach Rasmussen

Origin of Existence

Something Created From Nothing...

© 2012 Lulu Author. All rights reserved.

ISBN 978-1-300-30166-0

Acknowledgments

For the most part, I guess all I have to thank is you guys. Overall, in progress of writing this book, I have wrote it all am going to make my best attempt to keep it fully private until the book is finished. My overall goal is to be finished by October 15th, so about 2 weeks from writing this acknowledgments page. I began writing this book at the beginning of August, so if I finish on my set date I will have written this in about 2 1/2 months. Just wanted to say thanks for either ordering it on Lulu.com, Amazon.com or got redirected to one of those sites from TheTripleBstudios.com.

Introduction

I have always wanted to write a book of some sort. I know I will probably write this and be bored with writing books and not write again. But I am in the mood to do so and want to give you an introduction to me and this book. First an introduction to me : In case you don't know me, My name is Zach. I am a Youtuber and I am very interested in Astronomy, Cosmology, Physics, Philosophy, etc. and talk about it a lot. I, actually, at the moment am working on a documentary about the new Mars Rover as well. Now an introduction to this book : This book is going to give the "story" of everything in the universe. It is going to explain the beginning of the universe and the theorized end of the universe as well.

Zach Rasmussen

Origin of Existence

Something Created From Nothing...

Chapter 1

A Simple Perspective of the Universe

The Universe is a very complicated place and we are in the time where it first dawns on us on how it really works

I spend most of my time thinking about the Cosmos. The reality is having a deep interest in the Universe means that boredom is not something that happens often. Realizing that we are the product of the massive, ancient Universe to understand itself is pretty awesome. Understanding the 4 percent of the known Universe, our curiosity to know more and seek out the answers means that we are not allowed to be bored. Look at the cover of this book, the background is a snapshot from the Millennium Simulation (Go look it up on Youtube). The Millennium Simulation has been created due to the latest Super Computers. It is a simulation of something called "Super - Clusters". Super Clusters are basically large formations of galaxies in one massive area. The thing that blows my mind is inside one super cluster of many billions of clusters is home to an average galaxy called the Milky Way. Near the edge of that galaxy (26,000 light years from the center) holds an absolutely average star. That star has 8 planets in orbit : Mercury, Venus, Earth, Mars, Jupiter, Saturn, Uranus and Neptune. One planet, Earth, holds the only known kinds of life, which one species, humans, is intelligent. Although being intelligent beings with facts for all of this, a lot of people still live in the bronze age. A lot of people still believe in the talking snake, the burning bush and other things. I say that as if mocking them but honestly what makes know superiorly better? What makes my views somehow better than yours? Other than the facts, they aren't. Look, I wasn't there 14 billion years ago to witness the birth of the Universe, but neither was anyone else so why do we feel the need to make up a story to make us feel as if we know the theory of everything? For the creation of the Universe, you're guess is as good as mine... Unless we bring facts into the picture. I guess either way things like the big bang are impossible if the Earth is only 5,000 years old (joke intended). Other than that last joke, the fact of the matter is honestly I have no clue where to start. God of the gaps seems to be societies answer. If you are unaware, God of the gaps is taking all of the answers we don't know and replacing it with a supernatural being. I possess no desire to tell people what to believe, but things like that make people boring. Lets be clear, believing in God does not make you boring but saying you know everything about the unknown Universe when you obviously don't completely makes me think of you in a dull way. I heard an astronomer (I think it was Bill Nye) saying how if you replace all of the unknown with something made up you are boring and you have no curiosity anymore about the unexplored Universe. Without a doubt my all time favorite thing about cosmology and astronomy is looking at what I don't know and acknowledging that I don't know everything. It is much more interesting to want to know the answers than to make up fake answers and convince yourself you have the fully thought out M - theory (Theory of Everything). The bronze age held the belief of things like the Earth is flat, the Earth is the center of the Universe and constellations are celestial beings (literally). While believing the Earth was flat, it is quite clear that eventually one great mind will come along and show the truth, and they were Aristarchus, and Galileo. Lets start with Aristarchus, who was a greek mathematician and astronomer. Unlike others, Aristarchus was very fascinated by eclipses. Both solar, when the Moon passes in between the Sun and Earth, and lunar, when the Earth passes between the Moon and the Sun. Dating back to those ancient times whenever any form of an eclipse would happen, thats when God of the gaps comes back into the picture. That one I can almost see why they filled that in with a "God". Without a backed up scientific explanation about why that happened, I see why they would come up with some reason on the terrorizing moment when the sun vanished right before their eyes. Although Aristarchus was not fooled, he was certain there was a more logical explanation on why we have eclipses. He first noticed that they happen in patterns, so that had to account for something. Aristarchus also made complete and mathematical illustrations on this. This proposed that the Earth orbits the sun, and the Earth is round. He discovered the Earth is round due to eclipses in the Moon. The eclipses shadow on the Moon can only be the way they are if the Earth is a sphere. Aristarchus was also the one who made one of the most stunning realizations about the Universe and that is that the stars are just like our Sun, but only a very, very far way away. One major successor to Aristarchus is the famous Italian astronomer Galileo Galilei. Galileo was the biggest contributor to proposing the theory that the Earth is actually not the center of the Universe. Galileo was one of the founding fathers of modern day science. In 1610 Galileo became the first legitimate astronomer in history. He became the first human being to perfect the once thought idea of a telescope. It was on January 7th, 1610 where Galileo turned his telescope to the heavens. The first thing on Galileo's "to - do" list was to study the giant planet, Jupiter.

Galileo, so intrigued about the new way of studying the solar system, watched Jupiter every night. He first saw the giant gas planet and a few faint lights behind it. First he assumed the faint light were distant stars and ignored them. But the thing is they were not stars, but a few of the moons of Jupiter and they are now known as Io, Europa, Ganymede, and Callisto. On January 10th, 1610 he documented that one of them disappeared behind Jupiter. And a few days later it re-appeared on the

other side. Studying with a closer eye on the satellites (moons) he noticed that it kept happening and happening. He found that some objects in space are definitely not orbiting Earth as everyone had thought. Although, unlike what it would be today with a theory being accepted if given the proper evidence, no one accepted Galileo's theory on the center of the Universe that is not Earth. Galileo got in pretty big trouble with the Church, because the bible says that the Earth is flat, the center of the universe and does not move. Just for proposing the idea that the Earth is not the center of the Universe and orbits the Sun the Church wanted him to be executed (Doesn't that break one of the 10 commandments anyway?). Galileo admitted his "sin" and apologized about his heresy. Society decided to say that some objects just go on a complex route to orbit the Earth. Although apologizing about his theory and by dodging the bullet of missing execution, he was put on house arrest for the last 9 years of his life. For apologizing to the Church about his theory he still muttered "Eppur si muove" which basically translates to "But it does move..."

Chapter 2

A Doppler Shift

Before we get into the creation of the universe and the possible end, lets do an experiment :

Go outside and stand on the side of a street. Wait for a car to drive by, and you have heard this before, right? As it approaches the pitch of its engine rises and as it drives away the pitch of its engine drops. To put it simply its the sound a car makes when it drives by you when you are standing still relative to the car. It does that because as it approaches you the wavelengths get pushed together, making faster frequencies, therefore higher pitch. And as it drives away the wavelengths get stretched out, slower frequencies means lower pitch.

Now, Im assuming you were expecting this, but the same thing happens with light. You can't tell, but if we were to be watching the car, as it nears us the entire car would look to be slightly blue in color, and very slightly red as it drives away. It does that for the same reason sound does what it does : When it approaches the light particles/waves get pushed together, making a blue color. Then when it leaves the light particles/waves get stretched out, making a red color. Astronomers looking at galaxies through telescopes have noticed that far off galaxies are slightly red in color. Which means that they are moving away from us...

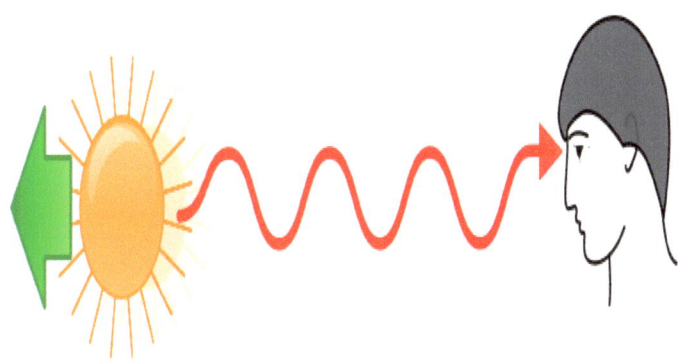

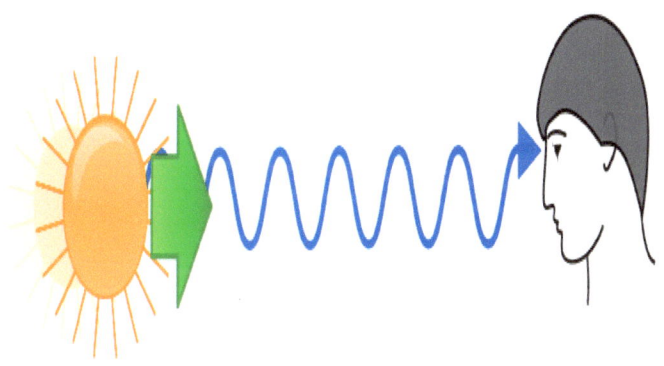

Chapter 3
The Big Bang Theory

Before the discovery of the big bang in the 1920s people thought that the universe was mostly static and unchanging. If you watch and observe carefully, you can come to the conclusion of a backed up theory with proof of how the universe began. You will also discover how unimaginably ancient the universe is : Roughly 14 *Billion* Years Old. Knowing that other galaxies are slightly red means that they

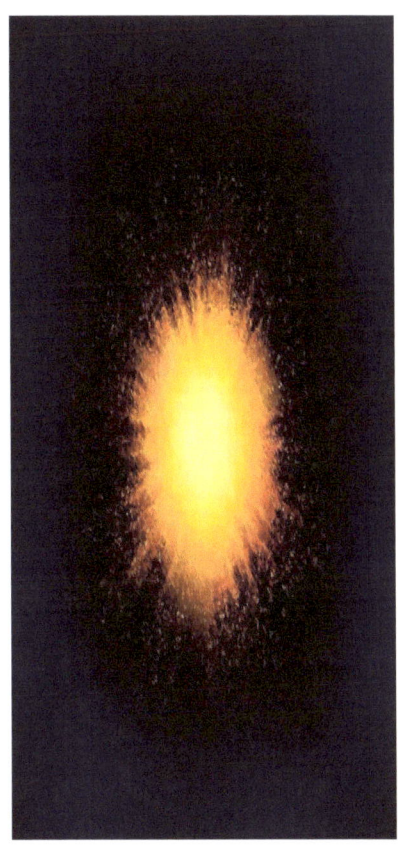

are moving away, which concludes that they have been moving away for us forever, so just take time, and make it run backwards. If everything is moving away from each other, then moving time backwards shows everything moving closer to each other. Do that enough you will see that everything moves closer and closer until the entire universe is one point of light. Then, the universe "exploded" into existence, which we now call the "Big Bang". In the moments of the big bang a force called "Dark Energy" was created. Dark Energy is the unknown force that accelerates the expansion rate of the Universe. But lets not think of the big bang of it as an explosion with fire and loud sounds. Because in reality the big bang was silent and in pitch black, because sound and light didn't exist at that point of time. It also wasn't an explosion in the sense of things being created and shooting in all directions, because it just began stretching and expanding.

But it raises the question : "What caused the big bang?". What made it start inflating? Nothing. I will explain how nothing made the big bang happen later but at least right now the universe is still expanding. For the moment a force called dark energy is making the expansion of the universe accelerating, its not what makes it expand, its what makes it expand faster and faster.

The big bang also happened very quickly, it started from smaller than an atom to the size of a baseball in less than a trillionth of a second, and about the size of our solar system in a minute and a half. It also, as it was expanding, became cooler, the small atom from what was the universe was basically a small point of heat, extreme heat. As it expanded it became bigger and cooler, and cooler. After that the creation of the big bang was a lot of matter, and its opposite : Antimatter. It was once theorized that matter should have a complete opposite of it, and it does. When matter and antimatter collide they destroy each other in one flash of light. They did that for a long time but luckily there was a little more matter than antimatter, if that wasn't so the universe wouldn't have anything in it. About 300,000 years after the big bang the fog finally cleared and everything became visible. So as far as we know thats how it all started, but we are only getting to the interesting part of the story.

Galaxies

About 1 Billion Years after the big bang the first galaxies to form. Our very own, Milky Way Galaxy is one of the oldest, forming 13 *Billion* Years ago, becoming one of the first. Galaxies need one object to be in the center which is the largest in the galaxy, for example, a black hole. Black holes are black voids in space, they are created when an old star collapses in on itself. There are small black holes, average black holes, then there are the ones in the center of the Milky Way : The *REALLY* big ones,

also known as "Super Massive Black Holes". Our galaxies black hole is roughly 11 million miles in diameter and is what everything in our galaxy orbits. It takes our solar system roughly 250 million years to orbit the center. Heres a weird fact about black holes : They aren't really black, they are invisible. You can't see a black hole, light can't escape from a black hole, therefore no light can reflect it. Fun Fact : Most people think that absolutely nothing can escape a black hole, but thats wrong, on thing can and it is called "Hawking Radiation". You can detect a black hole by the radiation it emits and the void it creates. Galaxies are one key factor for life, if you don't have a galaxy, you don't have a solar system, and no life.

The First Stars

Stars are pretty weird, the sun isn't as "normal" as we think. Stars are basically giant balls of hydrogen fusing together to create light and light in infra- red. Infra - Red is basically the heat, when you are cooking something in the oven and you open the door and that big gust of heat shoots towards you, thats infra - red. If either the light or infra - red didn't work we wouldn't have any stars, therefore no life.

Earth and the Moon

Earth was basically formed with matter orbiting each other and mashing up and creating the planet Earth. When it was first created it was a giant ball of magma. When it was created it was holding the left over matter of the stars and most importantly, the sun. So it was a very hot place, an 8,000 mile wide ball of heat created 4 billion years ago. It took Earth a long time for it to cool off for the sense for life to live on it. The moon, however, is a different story. There were many theories of how the moon came to be what it is today. The first theory is that Earth used to orbit much closer the the Sun, that meaning faster orbit and faster rotation. The faster Earth moved and faster rotation meant that the Sun's gravity actually yanked off part of Earth and formed the Moon. Another theory was that the moon was a lonely, drifting planet roaming around the galaxy until it was eventually caught in Earths orbit. That theory, however, has been proved wrong because the big size for the Moon would have meant it would have flown right past us. And the theory that is most accepted and probably right is that millions to billions of years ago Earth was a regular planet, and then a "Proto - Planet" about the size of Mars comes crashing into Earth. When that happened the collision made matter shooting out everywhere and most of the matter went into keeping Earth a planet, and the leftovers went into creating the moon we look up to today.

Chapter 4

Life

Thats how everything got its kick start, now we need life, preferably intelligent life to give the universe meaning.

How Did Life Start on Earth?

There are 2 main theories on how life began here on our small, pale, blue planet. The first is the most common and that is once our planet formed 4 billion years ago, scattered around were small pools of chemicals called "Amino Acids". Different cells would be floating around in those pools and would collide for millions of years, until we won the mega - lottery and the perfect combination for life just happened. That would be just organisms being made, but evolution is an amazing thing you know.

The other theory is that life could have been created on another planet, easier and more evolved for life, and would have been spread from planet to planet by astroids. Organisms would be carried deep inside the astroid and would be immune to the - 100 degree weather and the vacuum which is space.

Evolution

THE ORIGIN OF SPECIES

Complete and Fully Illustrated

CHARLES DARWIN

After we have organisms on Earth we have survival and evolution. Evolution is life adapting to new conditions, but it takes a long time. If its a whole new scenery they aren't used to, it can take 100 million years for them to fully adapt. But if its something small, like their eating habits, it can be only a few decades. It has taken a few billion years for organisms to become more advanced, like humans. Our oldest ancestor is actually a fish, while all life was living in the water. They evolved and moved out of the water and adapted to land-life. The Most primitive you can get with a species in which to evolve into other species which in the grand finally ends with us.

    But lets be clear how evolution works before we talk about how cool and convenient it is : When conditions aren't as good as they could be, animals go through evolution over a long period of time to adapt to the climate. I saw Christine O'Donnell say she thinks evolution is a myth because why is it when we go to the zoo that monkeys don't evolve before our eyes? *Sigh*... Lets break this down. First of all, that is implying that monkeys evolve into humans within a span of 20 minutes. But you also have to remember monkeys are living a pretty good life, there is no need for change. Monkeys live up trees and if you are living up a tree you probably would like to be a monkey. Even with evolution we have seen evolution doing what it does. A while ago, some evolutionists wanted to test evolution so they took a species of lizard and moved them to an unknown, unfamiliar island. It only took a matter of a few decades they had a whole new eating habit and their jaws were completely different along with their stomachs. Which evolution isn't even a crazy concept, it can

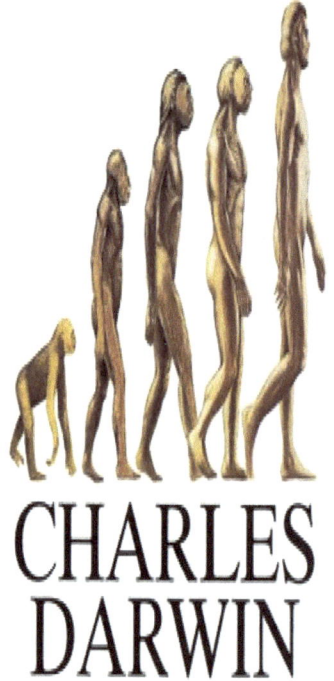

sound crazy the way you say it if you do say it like : "We came from monkeys." but the reality is : "When species experience unknown climate they adjust and adapt to it.". I know evolution is a controversial topic, but lets look on how it is true. Look at it like this : If evolution really is a myth, why do the facts say otherwise? Fact, reason and logic is how Charles Darwin discovered evolution and became the author of one of the most famous books "On The Origin of Species".

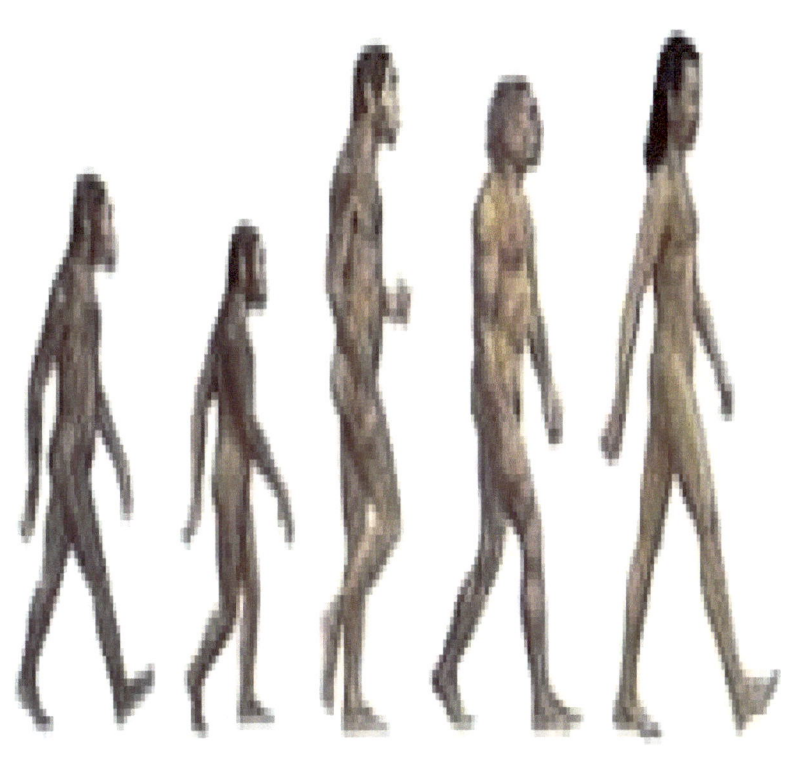

Darwin got the idea when he went on a long boat trip and saw the Galapagos islands. It also made him come to the realization that all animals adapt to climate and conditions over long periods of time. That we were organisms in the beginning evolved into monkeys, neanderthals, and humans, which debunks any notion of a supernatural being and shows that we don't even need a God. In the beginning people didn't exactly like the idea of evolution and not many people accepted it, but it wasn't to long after the publication of "The Origin of Species" that evolution was a very well accepted fact in the scientific community. I guess people don't like the idea of evolution because they think its "insulting", always knowing that we are superior to monkeys is kind of a turn off to hear that we used to be them.

    Darwin worked off the theory of evolution proposed by Lamarck. Lamarck proposed that evolution worked all in the hands of the parents of the offspring and that the traits the parents had would guarantee thats what their offspring would have. For example, he thought that if a fathers arm was cut off and then he "made offspring" that they would not have an arm. The problem with this is it has nothing to do with genes, the genetic code and DNA of the father does not include not having arms so their offspring would still have arms. Darwin's "Natural

Evolution of Planets and Galaxies

Not only do life forms go through the long process of evolution, but planets, stars and galaxies do to. The early stages of a planet forming is pretty dull, they are just hot rocks orbiting a star, it takes millions of years for the planet to cool down. This happens because the inner parts of the planet like the core have been recently formed and when they are new, they are hot, REALLY hot.

The Story of Humans

Humans, in comparison to the universe, have only been around for a small, small, small sliver of time. Generally humans have only been around for the last few thousand years. But that is just intelligent humans, neanderthals don't count in that scenario. We also have only had radio for the last few decades or so. Before humans there were Neanderthals, or cavemen, basically humans but with simpler brains. We, in an exploration point of view, have come very far. We started by leaving our home land in Africa roughly 20 million years ago. Once we became intelligent we explored on foot, then on boat, then on plane, now through the universe. We first started exploring space in the 50's and 60's. For NASA the Mercury missions and the Gemini missions were the first times of us exploring space. The Apollo missions were the first time we actually explored space with a particular goal. Apollo 1, unfortunately did not end up the way NASA expected it to. On January 27th, 1967 while running a simulation countdown before liftoff due to their directly fed oxygen tanks that are very flammable a fire sparked in the cockpit and killed all 3 of the astronauts inside. Apollo 4 and 7 were unmanned probes sent to the moon. Then there was Apollo 8, the first time a successful

Apollo mission took place. Apollo 8 did not land on the moon but did a full figure 8 orbit around it. Apollo 10 did basically the same thing but it was testing conditions for a landing. Then on July 16th, 1969 Apollo 11 launched and proceeded on July 20th, 1969 to have Neil Armstrong to be the first human being, to ever set foot on the moon. More landings like Apollo 12, 14, 15, 16 and 17 landed on the Moon. I skipped 13 because Apollo 13 never landed on the Moon due to an oxygen tank

explosion half way to the lunar surface and the astronauts had to turn around and come back to Earth. We, unfortunately, at the moment have to stop there. Mankind has not landed and walked on the surface on any other body than Earth for 40 years. The last time we walked on the moon was December 10th, 1972 on Apollo 17. That doesn't mean the age of space exploration is over, at the time I am writing this it is exactly one month since we landed a robot called "Curiosity" on the surface of Mars. Exploration for Humans : On the subject of space exploration, lets discuss the best exploration we have ever done. Voyager 1 was launched in August of 1977, now over 35 years old its journey has taken it to Mars, Jupiter, Saturn, and beyond the solar system. Voyager 1 is the only man made object that has left the solar system... technically. It is mostly out of the solar system because basically right now its in the Heliosphere, a magnetic bubble around the sun protecting us from massive amounts of galactic radiation. Voyager 1 is expected to be out of the heliosphere and into inter - stellar space by 2015. When NASA launched Voyager 1 and its companion Voyager 2 they were looking to explore and get a better idea of how big the solar system is, and the answer is really big. Using the gravitational pull and orbits of Jupiter and Saturn Voyager is blasting through space at 11 miles... per second. Yes, thats right, Voyager 1 moves through the universe at 11 miles a second, which is a full 40,000 miles per hour!

Is There A Meaning to Life?

While on the topic of life, we all ask the question wether it is simple organisms to intelligent beings like humans, is there a meaning to life? Is there a definite reason on why we exist at all? Philosophy was the first one to dawn on the question of meaning for life. I, for the most part, agree with professor Stephen Hawking when he says that philosophy is dead and the meaning to life can be explained with physics. I do agree that philosophy is dead and the meaning of life could be explained by physics, but just because philosophy has not kept up to date with modern science doesn't mean it doesn't hold the same key components. The meaning of life is relative, to me, meaning only exists on what we create in our minds. First lets look at it like this, next time you see you're family, you're friends, or someone you are in a relationship with and ask yourself : "Do I really think there is a meaning to life", if you're answer is yes, congratulations you've unlocked the key to meaning and reason. If you're answer is no, lets look at it again. Take the famous once thought experiment that goes : If a tree falls in the woods and no one is around to hear it, does it really make a sound? I know it will probably sound silly, but I am firmly on the side that it does NOT make a sound. Reason being that is that I view events to be relative, relative being it to me may not be relative being to you. Lets say I was around and heard the tree fall, in my universe the tree did make a sound, but to anyone who did not hear it, then in their universe the tree did not make a sound. So if no one was around to hear it then in every conscious being in the universe their perspective is nothing made a sound, so their must have been no sound. I also view it as that sound can only be sound once an ear has picked up the frequencies in the air and translated it to sound to the neurons in the human brain. This does relate back to the meaning of life. The surprising reality is the meaning of life is what purpose and meaning you put into life, is what you receive back. Our conscious minds are very complicated, if anything has meaning to you, then it has meaning to the universe. If you are the only being in the universe who thinks a certain rock you found has meaning, then it does have the amount of meaning to the world that you think it does. So if you look at you're life and ask if you're life has meaning to YOU, and you say yes, then it does, simple as that. So if the question is what is the meaning to life, then I will say that depends on who you are asking. Although it isn't that simple because there is one thing that perplexes us humans, and that is the theory of pre - determinism. Pre - determinism is the theory that no one has the ability of free will and when the Universe was created it set the timeline of what will happen and we, insignificant beings have no control over that. Most people deny this immediately because of the idea that we do not have free will is not worth even considering. Personally with though put into it, I believe we do have free will. Here are a few reasons, if there are alternate Universes with alternate realities what would it mean for them? If each alternate Universe has a different reality, from something big like there is no such thing of gravity or something small like instead of eating a hotdog one day you ate a hamburger. The idea of pre - determinism in here is to inconstant. If an alternate Universe is the exact same except for one point, 14 billion years in where instead of eating a hotdog for lunch you had a hamburger then it is not really an alternate Universe until that point. So pre - determinism could not have set 2 Universes to be identical Universes until 1 point. So they weren't "alternate" until 14 billion years in. Another reason on why I think pre - determinism cant work is because of choice. This one is also related to alternate Universe but lets start off with and example. Imagine you are faced with a choice : You can either go to the movies with your friends or you can go shopping by yourself. That is obviously what free will is, when our mind is faced with a choice we think about the options and choose one. If we were to choose one then theoretically an alternate Universe would spontaneously be created and fill in the other scenario. Next time you have a pair of dice roll them on the ground. Other than using enough mathematic predictions you cant make an educated guess on what it will roll. Same thing with the Universe, when the Universe was being created during the big bang the laws of physics were being set. Some may think that the laws of time (Pre - determinism) could have been set as well. But that is impossible because if pre - determinism was being set during the big bang, so during time, then there would have already been time being passed therefore no legitimate pre - determinism to have created a paradox. So in my opinion nothing can predict every move of the human brain. Some may think that means there was an intelligent creator who set time out from the beginning, but to prove that wrong with some thought out logic re - read the last few pages.

What is Reality?

Reality, we all know what it is... Or do we? I mean, reality with its best definition is the outside Universe when perceived by our brains. Our eyes can see light particles in the air and it perceives that and sends the report to a field of 100 billion neurons to convert that to something we view as something called reality. Before I go much further lets be clear about the fundamental difference : The brain is a collection of matter, cells and neurons. And the mind is our conscious being resulted from our working neurons. No matter what happens in the Universe, no one, absolutely no one will see the Universe the way you do. No one will see the world the exact way you do. That means 2 things : No one will see the exact same visual view that you do. And no one will philosophically see the Universe the way you do. So if everyone sees the Universe in completely different ways maybe there is no such thing as reality, no such thing at all... Well lets not jump to conclusions too quickly. Reality could change from being to being. What is reality could just be what is reality to me and reality to you combined. A scientific theory is that there is one state of mind of all human neurological thoughts being merged together. Although we are unsure if there is one single reality or not, just like the meaning of life there is a difficult theory we will have to think through a theory called "The Simulation Hypothesis". The simulation hypothesis is basically the theory that reality is not real and that our conscious minds are being run by a super computer. A futurist named Ray Kurzweil is known for predicting the idea of electric keyboard pianos before they were invented (thats why one of the most famous keyboard company is called Kurzweil). He also predicted the idea that in 2044 humans will be able to use supercomputers to stimulate our minds and make us think we are living a different reality, when it is simply a video game. If Kurzweil is right and that soon we can simulate an alternate Universe than that pretty much guarantees that reality is not real. But if there are fundamental limits to what we can realistically simulate, then that guarantees reality is real.

Is There An Afterlife?

For many people, this is common sense to think there is a Heaven, or any form of afterlife. For others, they just never think of it, the idea of death just isn't something that crosses their mind. I am on neither one of those, I actually do think about stuff like that a lot. Philosophical questions like that are what really interest me, my overall opinion on an afterlife is no. From my perspective, when you die you're brain simply shuts down, no more thinking, no more life. As Stephen Hawking said the human brain is like a computer, and will simply shut down when components fail. For whatever reason this is the topic that people get very superstitious about, so if you are one of those people skip past this. For the rest of you here lets take a deeper look into afterlife. Afterlife is the concept of when we die, our concuss soul will leave our body onto another dimension. I, for one, agree with professor Hawking on the matter of life is a matter of chance. I definitely think that life is an opportunity to be and appreciate to be the tool the universe uses to comprehend itself. Just remember, all of us, every single one of us is in fact the entire universe, living it as a human for a small period of time. I guess people believe in an afterlife for many reasons but the majority of it is because humans don't want to think they are 100% gone. I don't either, but it all comes down to : What do I want to believe? And what ***do*** I actually believe? Technically, our body will almost forever be in an alive state. Let me explain it like this, when you have a child, you pass you're genes onto them. So a lot of your cells go onto creating them. An interesting fact is about 1 billion of your cells, come from Shakespeare. Same thing with other people like Beethoven, Einstein, Darwin, etc. So a lot of your cells will forever be within your descendants.

Chapter 5

How it All Ends

Believe it or not modern day Cosmologists now have a pretty clear view of where Earth is going, where the Sun is going, where the MilkyWay galaxy is going, and most of all where the path of the universe is heading.

Despite some things I may sometimes say, I like to think of myself as an optimist, I am very hopeful and positive of where the human race is going. Don't get me wrong, I am realistic, just like every other species eventually humans will go extinct and leave Earth to be a lonely planet we used to call home. Or on the flip side humans may leave Earth behind and go to colonize the rest of our galaxy. Humans have one chance of surviving, and the answer is the Cosmos. I guarantee humans will go extinct if we can't escape our home planet we love so much, and rather than simply making about 5 lunar landings 239,000 miles away on the moon, we need to put our eggs in as many baskets as possible. We would be wise to be the pioneers of space exploration and / or colonization. Earth gets hit by a decently large astroid impact every few million years or so, so we definitely need to spread out as much as we can.

This, is Gliese 581D. Gliese is a very large, Earth - like planet and if humans do well, it will probably become home to our species. It orbits a smaller, redder star, but it is close enough that the conditions on it are just perfect for liquid water to exist on the surface. In the far, far future this planet could become home to the species called humans.

Although Gliese would be the perfect second planet for the Human race, there is a obstacle we definitely need to deal with. Gliese is an *Extremely, Extremely* far away planet. However it is the closest Earth - like planet Astronomers have yet to discover, but that doesn't mean its next door. For what could be our next home planet is still over 20 light years from Earth. Lets put that in perspective : A lightyear is how far light can travel in a year. Light moves at 186,000 miles per second, so moving at 186,000 miles per second for over 20 years straight could get us to Gliese. But Im afraid we have to stop a moment, even though that would still take 20 years the laws of physics dictate that it is physically impossible for a space ship (or anything that is not a photon) to catch up to the grand speed of 186,000 miles per second. But lets be clear : We can get very, very close to the speed of light, but we are incapable of getting to 100% of the speed of a photon of light. Although space ships could be capable of moving at about 99% of the speed of light, it still would take quite a bit longer than 20 years. Look at it like this : If the speed of light moves at 186,000 miles per second and 99% of the speed of light moves at, lets say 185,995 miles per second traveling for 20 years those miles would add up. Eventually the difference of 5 miles (which would be a few nanoseconds) would possibly end up adding a whole other decade.

Will Humans Survive?

A lot of people ask the question : Will humans survive? Will we get past our struggle? My answer : Possibly. The future of the universe is certain : The universe will continue on schedule for as long as it can no matter what humans do. But for the fate of our species, thats certainly uncertain. Humanity needs the kick to remind ourselves that there are over 50,000 nuclear weapons, and we can either use them to create a nuclear winter and kill all of us, or we can use it to colonize the universe that produced us. Some people are against NASA spending roughly 6 billion dollars in order to have the James Webb Space Telescope's launch and being put to work. My question is, why do we love war so much? Im not being sarcastic, completely serious. Look at it like this : It costs 6 billion dollars for the James Webb Space telescope, the new Hubble Telescope. For what is going to be the most advanced telescope being sent 1 million miles away from Earth to look deep into the past of the early universe and deep galaxies and planets and stars. You could do that, or you could spend the 6 billion dollars like we usually do, which goes into our American troops being in Iraq for 1 month, Its up to you. Our best chance is to spread out throughout space. We need to be as spread out as possible. If humans should be taken seriously through the universe, then we need to survive for millions, possibly billions of years. Earth is bound to be hit by an astroid once every few millions of years or so. The same thing that wiped out the Dinosaurs can be what will extinct the human race.

The Future and Our Place Within it.

The sun, a giant ball of hydrogen fused gases thats been lighting our solar system for the last 5 billion years. It is 93,000,000 miles away from Earth and 850,000 miles wide in diameter, and roughly 1,000,000 Earths can fit into the sun. Despite what I just said, the Sun is a perfectly average star in the mid corner of one of the oldest galaxies know, the Milky Way. The Sun is very small in comparison to the other stars like V-Y Canis Majoris, NML Cygni, etc. Our star, like most others, was created when enough hydrogen began fusing in the right place and due to one of the 4 forces of nature, gravity, forced it to form a star. In about 5 billion years, our star will basically die. It will haven heated up to 200 billion degrees. And you guessed it, it will not be peaceful and quiet. The Sun, running out of energy, will begin to expand, and expand to about 200 million miles in diameter. It will stretch out and eventually engulf our home planet we call Earth. A little after that point, the thing that used to have given us life has taken it away and shrunk down to a white dwarf. That would be the end of our solar system and the only known form of life. In some fortunate way humans could be the genuine pigs in intelligent life forms living in the Cosmos. In theorized reality, us humans, after we eventually go extinct, will go on to lead a future Cosmological generation. We pass on our genes to further species.

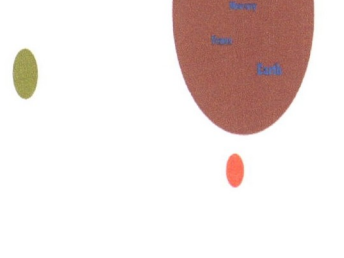

Our Solar System in 5 billion Years

Our very presence means we need to appreciate the Grand Design. Like I said earlier there is probably no Heaven, or afterlife at all. So that simply means we should appreciate what we have now, until its gone. Things like nuclear weapons and war is what is holding humanity back. The fact of life on Earth almost guarantees life exists somewhere else in the Universe, some intelligent species will thrive on to colonize the Universe. Others will produce nothing decent for themselves, and will go extinct and leave their planet to be dark, and cold. What will will we do with our fate? Will we use our technology to generate a nuclear winter and remove ourselves from the Cosmos? Or will we put our focus into our own benefit and colonize the unexplored? The choice is ours to make. I guess our knowledge and curiosity is what will make our decision. We can destroy ourselves and leave our planet to be empty, and dark, or we can expand our existence and knowledge of the origin of existence, its up to you.

Author

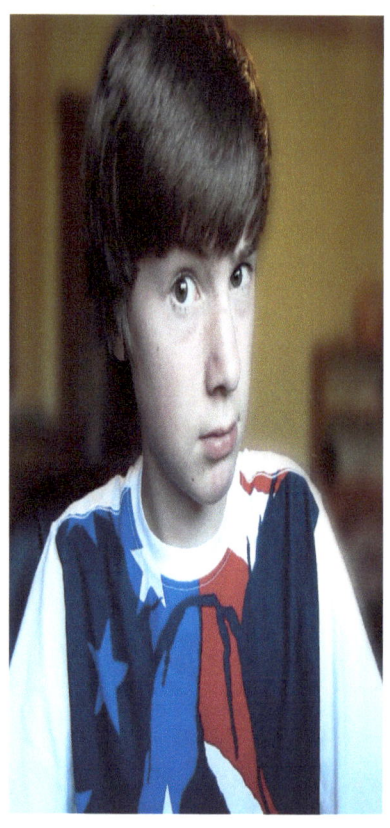

My name is Zach Rasmussen, I am a Youtuber, I guess a philosopher, Astronomy and Cosmology educator and now an author. I have been making Youtube videos for 2 1/2 years now and am loving it. I do primarily videos on the topic of Astronomy, Cosmology, Physics etc. I plan to write more books in the future because this has been one of the funnest things I have ever done. Thanks for reading :)

Bonus :

Here are a few more pages on some scientists that I really look up to :

Albert Einstein

There was once a physicist that believed matter can be a form of energy and vise versa, and his name was Albert Einstein. In 1905 Albert Einstein proposed his theory of relativity. The theory of relativity, or $E=MC^2$ was, and still is one of the biggest discoveries of science ever made. After many other equations he discovered the equation of $E=MC^2$. Lets break down the equation : E, which means energy, is equal to M, or matter. E=M but there is still M x C. C represents the speed of light squared (which is the 2). Adding that up you get the result of $E=MC^2$. He also discovered that matter actually drags on time. The closer you are to a very large object, like a black hole even though Einstein did not like the idea of black holes, time would slow down due to the great gravitational forces of the large objects gravitational field. Unfortunately Albert Einstein died during his search for the M-Theory (the Theory of Everything).

Isaac Newton

The idea of gravity was worked out by professor Isaac Newton, back in the 17th century. One of the 4 forces of nature, gravity, dawned on Isaac when he was sitting outside and an apple fell on his head. Seeing gravity in action helped professor Newton understand that large objects attract each other, and the larger the mass the stronger the gravitational pull.

Stephen Hawking

Without a doubt Stephen Hawking is the man I look up to the most. Not just in his work with physics and black holes, but as a person. But lets look at what he did for physics. Professor Hawking is very intrigued by black holes and in 1974 he proposed his equation on the matter of gravitational wells, his equation is $S_{BH}=kA/42P$. Basically that means that black holes with their unimaginable dense gravitational forces actually distort time. Hawking also added that black holes must emit radiation, as they do. So whenever someone says that nothing can escape a black hole correct them and teach them about Hawking Radiation. Stephen Hawking has also published multiple books, as my all time favorite book "A Brief History of Time". Stephen Hawking, when he was 21, was diagnosed with Motor Neuron Disease, other wise known as ALS or Lou Gehrig's disease. Its a form of muscular dystrophy that usually kills people within 1 year, but for Hawking he has lived with the condition for 50 years now, being the longest living human being with the disease.

Carl Sagan

Carl Sagan was an American Astronomer, Cosmologist, Author and science popularizer. He is best known for his PBS show he did "Cosmos : A Personal Voyage", the book "Cosmos", the novel "Contact", the Pale Blue Dot, SETI, etc. Sagan was the biggest popularizer of modern science, making it interesting for the world and my 2nd favorite scientist ever.

www.ingramcontent.com/pod-product-compliance
Lightning Source LLC
Chambersburg PA
CBHW041102180526
45172CB00001B/69